The StockMarketHedron

The Geometrical Jewel at the Heart of Finance

Ovidiu Racorean

To my daughter Anastasia

About the author

Ovidiu Racorean research work is focused on applied advance mathematical formalism to model the financial markets. Parts of the research expressed in pages of this book captured the attention of press and large public once an article in **Business Insider** called him "Hedge fund researcher working on a higher dimensional geometric model of the stock market".

Ovidiu Racorean is currently responsible of the quantitative research unit in a **hedge fund**. Due to the confidentiality policy at his work place some aspects related to research presented in the book remain undisclosed for now.

Preliminary Remarks

The most advanced mathematical formalism that leads scientists working in Quantum Field Theory, to discover a new geometrical shape-the **Amplituhedron**- is used, to reveal, step-by-step the stock market geometry.

The mathematics lying at the heart of the **Universe** is brought down on **Wall Street** to predict stock market price movements.

The StockMarketHedron is not a "crystal ball", but a jewel-like geometrical object that encodes all the relevant information on the stock market. It helps stock market professionals to choose the most probable future state for their investments.

The reader is invited throughout the pages of the book to find the answer of the question:

IS *DOW JONES INDUSTRIAL AVERAGE* A GEOMETRIC-LIKE OBJECT?

Author and his work in press releases

"Hedge fund researcher is working on a higher dimensional geometric model of the stock market"- **Business Insider**

"How do you picture the stock market: a bunch of guys yelling at computer screens on Wall Street? A long list of figures in the paper? Or, perhaps, an ever-shifting, higher dimensional jewel? The latter vision is that of Ovidiu Racorean... " **- International Business Times**

Unbelievable！研究发现预言股市的"水晶球" -
Laohucaijing

Articles in full and complementary materials related to the subject can be found at:

www.quantumfinancialmarkets.wordpress.com

Table of contents

Book overview

A general stock market index is a method to measure the state of a market evaluating the prices of a selection of stocks owned by the most representative companies. These selected stocks are the components of the index. The market index is computed typically by a weighted average of the selected stocks prices and the performance of stock market can be asses.

Such a global view of market performances, although extremely valuable, is not offering an image of what happen with the stock components in the evolution of the market quotations. Viewing the relations between stock components of a market index on a daily basis gave valuable complementary information that help in the investment decision making process.

Prices of the stock index components can be arranged in a table in ascending order starting from smaller stock prices, in the left, to companies having higher stock prices, to the right. Section 1 is devoted to explain the concept in detail, here is suffice to underline the main reason for such stock order. It could be noticed that along the time some stocks surpass or come under the price of neighbors stocks in the parallel series. Other way to see this is to imagine the stocks are crossing their price series.

To exemplify the stocks crossing diagram with real stock market prices, the market index is chose to be *Dow Jones Industrial Average* (DJIA) with its components. Prices of a fraction of all 30 components of DJIA are arranged in the manner explained above, starting from CSCO which is the lowest priced stock until the highest, PG, for the market reference date 5/15/2013.

To simplify the understanding of these particular diagrams only 4 stocks are retained and the prices wiring diagram is drawn, along with relevant examples of stock crossings.

This simple technic of viewing the stocks prices of the DJIA components reveals the deep connection that exists between stocks crossings and permutation diagrams. It can be said that every time the price of a stock in an index go up or under its neighbor stock price that the two stocks permute.

A deeper insight into measuring the performances of a market by examining the permutations of the stock components of a representative index on that market can be possible by building a bridge that directly connects stock markets to more complex structures that exist in combinatorial mathematics, such as decorated permutations, hook diagrams or positroids.

Sections 2 will constitute an insight on the last years work of the mathematician Alex Postnikov related to connections between wiring diagrams , decorated permutation, and hooks. Examples on how these beautiful combinatorial objects are related to the states of the stock market are provided.

The last bricks in the construction of the bridge relating stock market to geometry behind it are shown in Section 2. New combinatorial structures are coming into the "scene" and are explained in association with financial concepts, in their order of "apparition". The remarkable structures of matroids, positroids are effectively connected to price quotations of stocks in the market, such that a new geometrical object emerges, the *stock market polytope*.

Polytopes are geometrical structures that live in higher dimensions such that is typically impossible to represent them graphically. The example of the chosen four DJIA components is exceptionally provided a beautiful polytope structure depicted at the end of this section.

Section 2 is a refinement of the combinatorial approach to stock market, intermediated by Grassmann geometry. The reader is briefly exposed to the more complex geometry of *positive Grassmannian* , intensively studied in the latter years, mainly due to many applications of it in Quantum Field Theory. Technical terms like *positive Grassmannian cells* and positroid decomposition may look complicated, but are easy to be explained as removals of some stock crossings, and help in deeper understanding the properties of stock market polytope.

Considering the DJIA example of a market index, it is straightforward to see that because of the large number of components, the stock market polytope associated to it will be complicated and that it won't be easy to compute. Still, using dedicated software like MATEMATHICA it should relax the process to reveal the shape of the usual stock market polytope or to define geometrical structure of market under stress, like the 2008 crisis.

The volume of the stock market polytope can be determined and will evaluate the probability of a certain state of stock market, but this will be the subject of a future research.

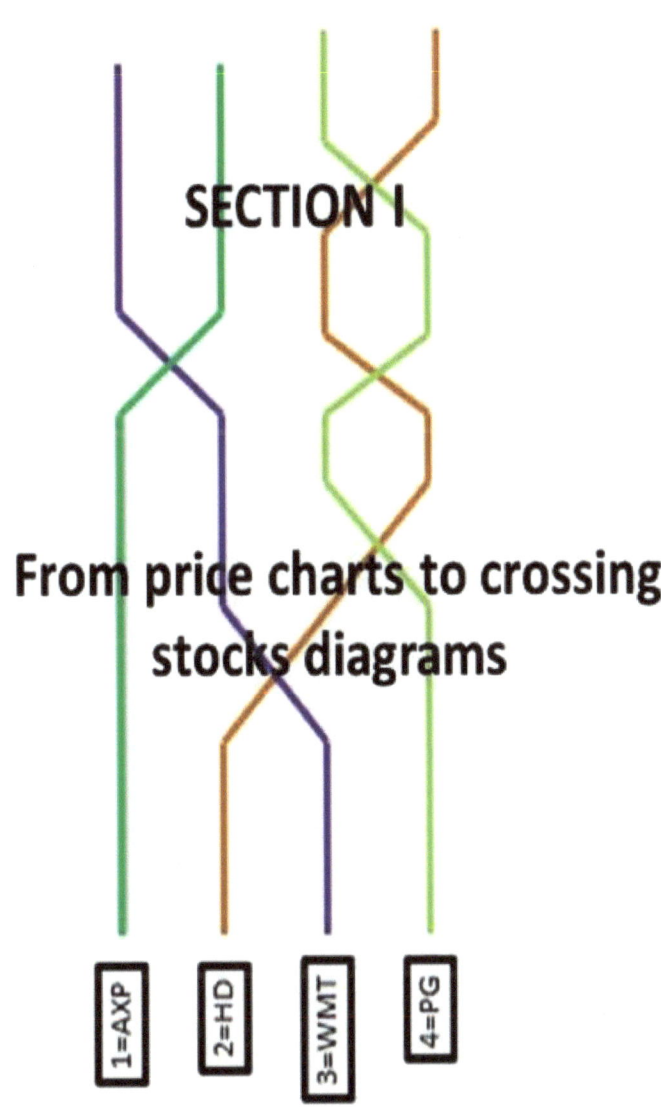

SECTION I

From price charts to crossing
stocks diagrams

1=AXP 2=HD 3=WMT 4=PG

SECTION I: From price charts to crossing stocks diagrams

In a chart representation, price of a stock forms patterns almost everybody is familiar with. When it comes to analyze a market index like **Dow Jones Industrial Average** (DJIA), all the stock components are represented in the same chart in a cumulative time series. In such scenario the paths of stocks prices are clearly crossing which other from time to time as the price of a certain stock gets over /under the price of an adjacent stock.

A surprising image of the stock market arises considering the price time series of all *Dow Jones Industrial Average* (DJIA) stock components only in their interactions in the form of the crossings. The chart of cumulative time series evolves in a **crossing of stocks diagram**.

The crossing stocks diagram method is capable to uncover hidden remarkable geometrical and topological properties stock market is endowed with, properties that will be largely explored in the next sections.

Chapter 1. Briefing of a general stock market index

The idea to measure the average performance of the stock market as a whole is old and can be traced back in the mid-1880 when Charles Dow, Edward Jones and the journalist Charles Bergstresser launched for the first time the concept of market index. The three analyst pioneers averaged the prices of the most influential nine companies stocks in their first attempt to obtain a global image of the stock market in the form of a market index. In 1896 *Dow Jones Industrial Average* index was born in its own rights.

A general stock market index is a method to measure the performances of a market by a selection of the most representative stocks that are traded on the respective market. It is computed typically by a weighted average of the selected stocks prices. The levels of the market index indicate the state of the market, a market in contraction for small index levels or a market in expansion for high levels of the index.

Such a global view of market performances, although extremely valuable, is not offering a complete image of what happen with the stock components in the evolution of the market quotations, how the relations and correlation between them changes along the time. Some attempts to reveal the relations between market index stock components were made by T. Preis and co. and covered particular periods in the market existence such as the 2007 financial crisis, see for details [11]. This work uncovers some hidden correlations between stocks in financial crisis period and highlights the importance of accounting the influence of all stock components in the behavior of the market index.

Although, viewing the relations between stock components of a market index on a daily basis is not crucial for market professionals, it gave valuable complementary information that help in the investment decision making process. The present book constitutes a unique and completely different attempt to offer investors rapid images of the market moves. In the following sections a map relating stock market with complex geometrical and topological objects is created, such that at the end of the presentation any trader and investor would be capable to interpret the market moves only by seeing geometrical or topological shapes behind stock market transactions.

	CSCO	GE	INTC	PFE	MSFT	NKE	UNH	AXP	HD	WMT	PG
6/7/2013	24.5	23.9	24.6	28.3	35.7	62.8	62.6	78	78.7	76.3	77.8
6/6/2013	24.6	23.4	24.7	28.1	35	62.2	61.9	76.2	77.3	75.6	76.8
6/5/2013	24.3	23.3	24.7	27.5	34.8	61.8	61.8	74.8	75.1	75.3	76.7
6/4/2013	24.4	23.7	25.4	27.7	35	62.8	62.4	76.1	76.6	75.9	77.4
6/3/2013	24.4	23.6	25.2	27.8	35.6	63	62.8	76.5	79.1	75.7	77.7
5/31/2013	24.1	23.3	24.3	27.2	34.9	61.7	62.6	75.7	78.7	74.8	76.8
5/30/2013	24.4	23.6	24.2	28.3	35	62.4	64.7	76.1	79.4	75.6	79.1
5/29/2013	24.1	23.6	24.3	28.3	34.9	62.9	63.4	75.8	79.5	76.2	78.9
5/28/2013	23.9	23.6	24.1	29	35	63.3	63.3	76.2	79.8	77.3	80.9
5/23/2013	23.5	23.7	24.1	29.1	34.2	63.3	62.4	74.7	78.9	76.3	78.7
5/22/2013	23.3	23.9	24.1	29.3	34.6	64.5	62.3	74.4	79.7	77	78.8
5/21/2013	24	23.7	24.2	28.8	34.9	65.2	62.9	75.1	78.7	77.4	78.8
5/20/2013	24	23.6	24.1	28.7	35.1	65.3	62.6	74.4	76.8	77.4	79.1
5/17/2013	24.2	23.5	24	29	34.9	65.3	62.8	73.3	76.9	77.9	80
5/16/2013	23.9	23.3	23.9	29.3	34.1	64.4	62.1	72.2	76.8	78.5	80.2
5/15/2013	21.2	23.2	24.2	29.6	33.9	65.8	61.6	72.8	77.9	79.9	80.7

Table I.1. A fraction of the DJIA index components sorted by price quotations from the left to the right at 05/15/2013

To simplify the discussion and give examples from real stock market conditions, and also, to celebrate its long existence, the index of interest is choosing to be the *Dow Jones Industrial Average* (DJIA). A fraction of price quotations for some DJIA components are shown in table I.1 as daily closing prices for a period in 2013 between 5/15/2013 and 6/7/2013.

It can be easily seen that the DJIA stock components in the table I.1 are arranged in ascending order from the stock with the smallest price quotation (CSCO) at the left to the stock with the highest price (PG) at the right at the start date 5/15/2013. This will be a rule of arranging the stock components of the DJIA index.

Notice that for others rows of the table this rule is not applied, such that next day, at 5/16/2013, for example, the price of CSCO, 28.9 came over the price of GE, 28.8. The next section will explore the consequences of arranging all the rows in the table containing the stock components of DJIA in ascending order from the left to the right and some unexpected and beautiful combinatorial results will be revealed.

Chapter 2. From price charts to crossing stocks diagrams

How to analyze the price time series of market index stock components all at once? Since the easiest and familiar way of analyzing stocks prices is the **chart**, it seems that the answer is to arrange the time series of all market index stock components in **one chart**, as a cumulative time series chart.

A fraction of price quotations for four DJIA components, AXP, HD, WMT and PG are shown in figure I.1 as daily closing prices for a period in 2013 between 5/15/2013 and 6/7/2013, along with the chart presenting the time series of all 4 stocks.

	AXP	HD	WMT	PG
6/7/2013	78.04	78.74	76.33	77.75
6/6/2013	76.24	77.26	75.63	76.82
6/5/2013	74.76	75.1	75.25	76.66
6/4/2013	76.06	76.63	75.94	77.37
6/3/2013	76.47	79.08	75.69	77.66
5/31/2013	75.71	78.66	74.84	76.76
5/30/2013	76.14	79.44	75.63	79.09
5/29/2013	75.83	79.49	76.23	78.9
5/28/2013	76.16	79.82	77.32	80.86
5/24/2013	75.27	78.99	77.31	81.88
5/23/2013	74.69	78.91	76.33	78.7
5/22/2013	74.44	79.69	77.03	78.82
5/21/2013	75.11	78.71	77.39	78.8
5/20/2013	74.4	76.76	77.4	79.09
5/17/2013	73.32	76.86	77.87	80.02
5/16/2013	72.23	76.75	78.5	80.2
5/15/2013	72.78	77.88	79.86	80.68

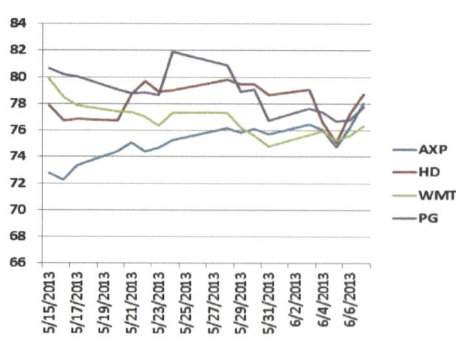

Figure I.1. Representing all stocks prices time series in one chart at once, as a cumulative time series

It can be seen from the chart in figure I.1 that trajectories of stocks prices are crossing and re-crossing each other as one stock price exceed or came under the price of an adjacent stock, in other words every time the **stocks are crossing**. The most important concept residing from this particular arrangement of the market index stock components is the **crossing of stocks.**

2.1 Method 1 – interpreting price charts as crossing stocks diagrams

Following the colored stock time series in the chart above a stocks crossing diagram can be drawn simply by straightening the price paths of stocks for the periods where no crossings exists. The *crossing of stocks diagram* explicitly shows the moments when the price of a stock comes over or under the price of its neighbor stock as is shown in figure I.2.

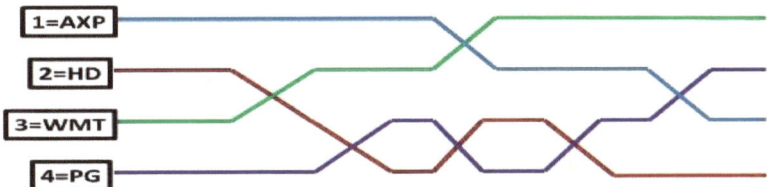

Figure I.2. Crossing of stocks diagram, a stylized representation of the chart in figure I.1

Figure I.2 replicates the chart in figure I.1 consisting in all stocks time series showing only the crossing of stocks. Notice that it cannot be said from the crossing diagram which is the stock that have the price overcoming or under coming the price of its adjacent stock. In the chart in figure I.1 it can immediately notice that in the interval from 5/20/2013 to 5/21/2013 the price of HD came over WMT and PG, even only by observing the graph of every stock. Section III will be devoted to diagrams incorporate this important result.

2.2. Method 2 - Arranging the stocks prices in colored tables

It can be easily seen that the DJIA stock components in the table I.1 are arranged in ascending order from the stock with the smallest price quotation (CSCO) at the left to the stock with the highest price (PG) at the right at the start date 5/15/2013.This will be a rule of arranging the stock components of the DJIA index.

Notice that for others rows of the table this rule is not applied, such that next day, at 5/16/2013, for example, the price of CSCO came over the price of GE.

To simplify the exposition only four stock components of DJIA are retain further, AXP, HD, WMT and PG. The number of stocks is chosen such that the discussion should be neither trivial, nor too complex.

The price quotations for the chosen four DJIA components are arranged in ascending order from the stock with the smallest price (AXP) on the left to the stock having the highest price (PG) on the right at the starting date 5/15/2013. The time series of prices for every stock is colored in a different color as is shown in figure I.3 a). The arrangement of stocks from the left to the right in ascending order of prices is preserved for every row in the table, say for every trading day. In this manner the stocks prices will be shifted from their initial positions (see figure I.3 a), at the right or left, every time the price of one of the four stocks comes under or over the price of the neighbor stock, put it in other words every time the **stocks are crossing**. Figure I.3 b) depicted the sorted prices of stocks in ascending order from the left to the right, and the crossings of stocks become very clear by following the stocks colors.

	AXP	HD	WMT	PG		AXP	HD	WMT	PG
6/7/2013	78.04	78.74	76.33	77.75	6/7/2013	76.33	77.75	78.04	78.74
6/6/2013	76.24	77.26	75.63	76.82	6/6/2013	75.63	76.24	76.82	77.26
6/5/2013	74.76	75.1	75.25	76.66	6/5/2013	74.76	75.25	75.1	76.66
6/4/2013	76.06	76.63	75.94	77.37	6/4/2013	75.94	76.06	76.63	77.37
6/3/2013	76.47	79.08	75.69	77.66	6/3/2013	75.69	76.47	77.66	79.08
5/31/2013	75.71	78.66	74.84	76.76	5/31/2013	74.84	75.71	76.76	78.66
5/30/2013	76.14	79.44	75.63	79.09	5/30/2013	75.63	76.14	79.09	79.44
5/29/2013	75.83	79.49	76.23	78.9	5/29/2013	75.83	76.23	78.9	79.49
5/28/2013	76.16	79.82	77.32	80.86	5/28/2013	76.16	77.32	79.82	80.86
5/24/2013	75.27	78.99	77.31	81.88	5/24/2013	75.27	77.31	78.99	81.88
5/23/2013	74.69	78.91	76.33	78.7	5/23/2013	74.69	76.33	78.7	78.91
5/22/2013	74.44	79.69	77.03	78.82	5/22/2013	74.44	77.03	78.82	79.69
5/21/2013	75.11	78.71	77.39	78.8	5/21/2013	75.11	77.39	78.71	78.8
5/20/2013	74.4	76.76	77.4	79.09	5/20/2013	74.4	76.76	77.4	79.09
5/16/2013	72.23	76.75	78.5	80.2	5/16/2013	72.23	76.75	78.5	80.2
5/15/2013	72.78	77.88	79.86	80.68	5/15/2013	72.78	77.88	79.86	80.68

Figure I.3. I.3.a) The initial arrangement of stocks **I.3.b)** The prices of stocks are sorted showing the crossing of stocks

13

Notice from the figure I.3 b) above that from time to time *stocks are crossing.* As an example, it can be seen that at 5/21/2013 the closing prices of HD and WMT are crossing. As it was stated earlier the impact of the crossing on the total value of DJIA is, for now, neglected and the attention is focused only on the colored trajectories the stocks prices take in the DJIA components table.

Bearing in mind only the colored trajectories of stocks prices, totally neglecting the values of quotations, the figure I.3 b) can be transpose in the picture bellow, where the crossings of stocks become very clear:

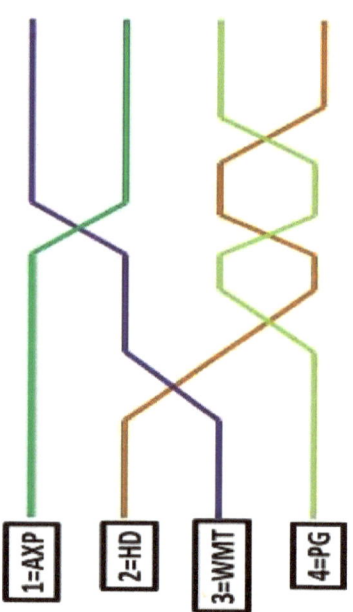

Figure I.4. Crossings of stocks diagram

The **crossing of stocks diagram** although simple, hidden remarkable mathematical proprieties which will be explored in the next sections.

Conclusions of the section

The present book proposes a way of interpreting the behavior of market indexes as a multivariate time series that results by arranging the index stock components in a simply particular manner. Prices of the stock index components are arranged in a table in ascending order starting from smaller stock prices, in the left, to companies having higher stock prices, to the right. Coloring every stock prices time series in different a color is the way of keeping track with each stock price quotations in the arranged table.

Although simple, this technique of arranging the stocks encodes immense mathematical potential and unlocks some remarkable geometrical and topological objects such as permutations, polytopes, braids and knots, to show up in the financial frame.

The most important concept reside from this particular arrangement of the market index is the **crossing of stocks.** Following the colored stock time series in the arranged table a stocks crossing diagram can be drawn. The ***crossing of stocks diagram*** explicitly shows the moments when the price of a stock comes over or under the price of its neighbor stock in the table.

To exemplify the stocks crossing diagram with examples from the real stock market, the market index is chose to be *Dow Jones Industrial Average* (DJIA) with its components. Prices of a fraction of all 30 components of DJIA are arranged in the manner explained above, starting from CSCO which is the lowest priced stock until the highest, PG, for the market reference date 5/15/2013. To simplify the understanding of these particular diagrams only 4 stocks are retained and represent the main constituents to exemplify the new mathematical concepts at.

SECTION II

Geometry of the Stock Market

SECTION II: Geometry of the Stock Market

It seems to be very unlikely that all relevant information in the stock market could be fully encoded in a geometrical shape. Still, the present paper will reveal the geometry behind the stock market transactions. The prices of market index (DJIA) stock components are arranged in ascending order from the smallest one in the left to the highest in the right. In such arrangement, as stock prices changes due to daily market quotations, it could be noticed that the price of a certain stock get over /under the price of a neighbor stock. These stocks are crossing. Arranged this way, the diagram of successive stock crossings is nothing else than a **permutation diagram**. From this point on the financial and combinatorial concepts are netted together to build a bridge connecting the stock market to a beautiful geometrical object that will be called **stock market polytope**. The stock market polytope is associated with the remarkable structure of *positive Grassmannian* .This procedure makes all the relevant information about the stock market encoded in the geometrical shape of the stock market polytope more readable.

Chapter 1. Crossing stocks and permutations

Recalling the grossing of stocks diagram in figure I.4 and removing the "bubble" formed in the diagram from 5/24/2013 to 6/5/2013 that has no influence on the stock market evolution, as it will be explained in the next sections, a simplified crossing of stocks diagram can be depicted in the figure II.1 below:

Figure II.1. Simplified form of crossing stocks diagram (wiring diagram)

The resulting diagram as appeared in the figure II.1 is known in combinatorial mathematics as **_wiring diagram_** and is associated to permutations. A **permutation** is simply related to arranging and rearranging of a certain set of values, as is the case with the price of stocks that compose a market index.

21

The permutations are noted $\pi = \begin{pmatrix} 12..n \\ \pi_1\pi_2...\pi_n \end{pmatrix}$ or simply

$\pi = \{\pi_1, \pi_2, ..., \pi_n\}$ and states that a set of n elements is

arranged in π_n way. Focusing the attention on the wiring diagram

II.1, the permutation associated to it is $\begin{pmatrix} 1234 \\ 2413 \end{pmatrix}$ or $\{2, 4, 1, 3\}$,

and simply says that 1 goes in 2, 2 in 4, 3 in 1, and finally 4 goes in 3.

It is easy to interpret this permutation in terms of the stocks in the figure II.1. Noting 1=AXP, 2=HD, 3=WMT, 4=PG, the permutation $\{2, 4, 1, 3\}$ simply states that for a period of time from 5/15/2013 to 06/03/2013 price of the stock AXP permuted with the price of the stock HD, HD permuted with PG, WMT with AXP, and finally WMT change the place with PG.

The crossings of stocks appear in a different light now, it is easy to notice that are nothing else than permutations of stocks.

The next chapter will explore the fascinating world of combinatorics and the ways that beautiful combinatorial objects are related to changings of stocks prices quotations in the market.

Chapter 2. "Decorating" the stock market permutations and some combinatorial remarks

Wiring diagrams associated to stock permutations as the one in figure II.1 have a beautiful visual impact. Still, in order to extract important combinatorial aspects related to stock arrangements, further mathematical tools are necessary.

The stocks permutations depicted in the wiring diagram in figure II.1, written as $\{2, 4, 1, 3\}$ can be represented in a more simply way, named permutation diagram:

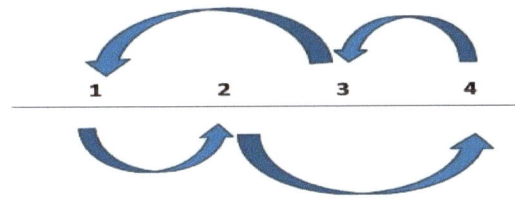

Figure II.2. Permutation diagram for a selection of four DJIA stock components

The permutation diagram in figure II.2 above encodes the same information as wiring diagrams. Notice that the permutation graphic has two arcs pointed to the left and two arcs pointed to the right. The permutation diagram has the advantage that it can be extended to combinatorial objects, such as *decorated permutation* and *Grassmann necklace*, which will eventually leads to geometric representation of stock market.

Decorated permutation is a permutation $\pi = \begin{pmatrix} 1 \, 2 \, ... \, n \\ \pi_1 \pi_2 \, ... \, \pi_n \end{pmatrix}$

with fixed points $\pi_i = i$ colored in two colors.

This "dry" definition of a combinatorial object is actually hiding some common sense situations normally encountered if evaluate the components of a market index at different periods. In the market evolution of stock prices there are stocks that do not cross with the other nearest stocks. Not having a transposition, for these stocks the points in the permutation are fixed, so that $\pi_i = i.$

Let's take a look again at the figure II.1 and analyze the permutation that result indexing the stocks quotations at 05/29/2013. The associated permutation, in this case, is $\{1, 3, 4, 2\}$. It can be noticed that stock 1 (AXP) do not cross with its consecutive neighbor (HD) and remain at its initial place, in other words AXP is a fixed point, and $\pi_1 = 1.$

In the same register, resuming the market quotations at 06/05/2013 the associated permutation becomes $\{1, 3, 2, 4\}$. This time, both stock 1 (AXP) and stock 4 (PG) are fixed points. These two fixed points must have cords associated in the permutation diagram and for that the following convention will take the place of the coloring function in the decorated permutation:

o   the cord associated to a stock having the price increased from the initial moment is pointed to the right;
o   the cord associated to a stock having the price decreasing from the initial moment is pointed to the left.

In the permutation $\{1, 3, 2, 4\}$ there are two fixed points, 1 and 4 associated to stocks AXP and PG. The price of stock AXP increase from 72,78 at 05/15/2013 to 74,76 at 06/05/2013 and its cord points to the right; the price of PG decrease in the same period from 80,68 to 76,66 and the cord associated to it will points to the left.

The permutation diagram associated to permutation $\{1, 3, 2, 4\}$ is depicted in figure II.3 and is describe a decorated permutation.

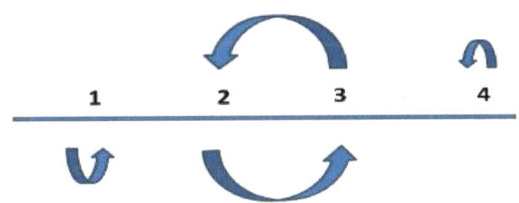

Figure II.3. The decorated permutation $\{1, 3, 2, 4\}$

Decorated permutation is in bijection with another remarkable combinatorial object, Grassmann necklace. The Grassmann necklace makes the connection with geometry that will be induced to stock market. As before, the combinatorial definition will be provided and the relation with stock market will be revealed.

A **Grassmann necklace** is a sequence $I = (I_1, \ldots, I_n)$ of subsets $I_i \subseteq \{1, \ldots, n\}$ such that

- if $i \in I_i$ thn $I_{i+1} = I_i \backslash \{i\} \cup \{j\}$ for $j \subseteq \{1, \ldots, n\}$;
- if $i \notin I_i$ then $I_{i+1} = I_i$.

It will be soon noticed that it is easy to find the Grassmann necklace from a decorated permutation despite the raw combinatorial definition above. Every term I_n in the Grassmann necklace sequence I is formed by the cords pointing to the left in every cyclically shifted ordering of permutation terms.

The Grassmann necklace concept is depicted in figure II.4 where every term in the cyclically shifted order is computed as being a part of the set I. The decorated permutation $\{2, 4, 1, 3\}$ is the same as before, and is based on the stock market permutation associated to the wiring diagram in fig II.1.

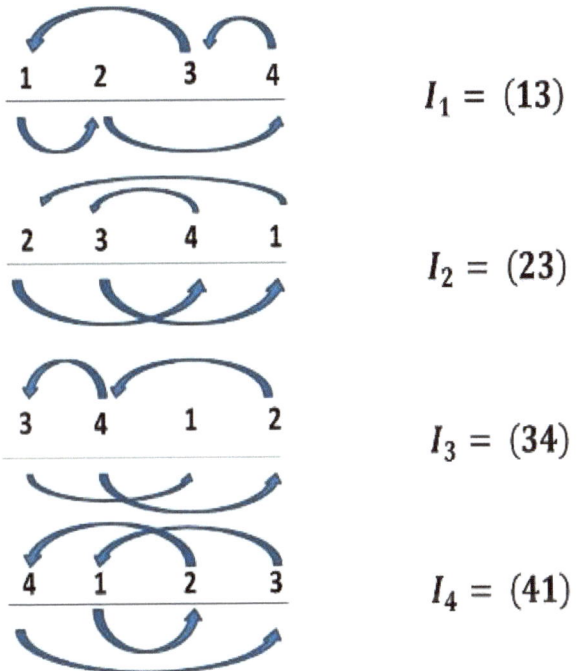

$$I_1 = (13)$$

$$I_2 = (23)$$

$$I_3 = (34)$$

$$I_4 = (41)$$

Figure II.4. The terms of the Grassmann necklace associated to permutation $\{2, 4, 1, 3\}$

It is easy to notice that are two cords pointing to the left and that will make every term in the Grassmann necklace to have two components.

Just for exemplification consider the Grassmann necklace terms as being expressed by the stocks:

(II.1)

$$I_1 = (AXP \ WMT), I_2 = (HD \ WMT),$$
$$I_3 = (WMT \ PG), I_4 = (PG \ AXP).$$

Another way to construct the Grassmann necklace is explored in the next chapter by means of hook diagrams which constitutes an even richer object in aspects related to combinatorial concepts.

Chapter 3. Hooks diagrams for stock permutations

Hooks diagram not only help to find Grassmann necklace of a decorated permutation, as was pointed out in the last chapter, but also, more important, will provide information on the number of dimensions the final stock market geometrical object will have. The geometry of the stock market for the chosen four stock components of DJIA will be 3-dimensional. Considering all 30 stock components the geometry will be far more complicated, and the resulting geometrical object will live in a space with many more dimensions.

To construct the hooks diagram an alternative way to describe the decorated permutation is necessary. The decorated permutations are viewed this time as a map $\pi : \{1, ..., n\} \rightarrow \{1, ..., 2n\}$ such that $i \leq \pi_i \leq i + n$. Simply speaking the cords pointing to the left in a decorated permutation are sending beyond the initial n terms of the permutation, that means π must be shifted by n relative to the initial permutation.

Turning to the example of the DJIA four stocks components, the permutation $\{2, 4, 1, 3\}$ associated to the stock market must be shifted to $\{2, 4, 5, 7\}$.

Saying all this and skipping, for the sake of simplicity, the mathematical definitions, hooks diagram associated to the above permutation can be depicted graphically as:

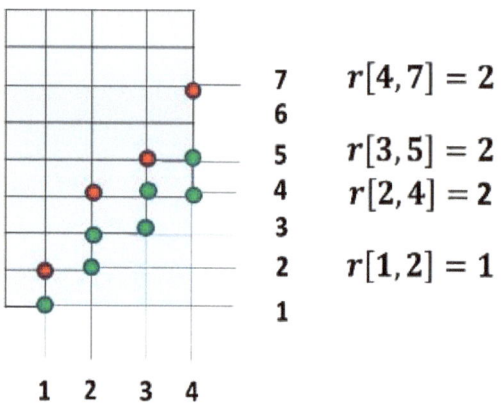

$$r[4,7] = 2$$

$$r[3,5] = 2$$
$$r[2,4] = 2$$

$$r[1,2] = 1$$

Figure II.5. Hooks diagram associated to permutation $\{2, 4, 1, 3\}$

Interested reader in construction of hooks out of decorated permutation associated to it, is referred to [3], [32], [33] and [34].

The hooks diagram encodes the important information about the dimensionality of the geometrical object that stock market is associated with, through the decorated permutation.

In the diagram above the $r[i, \pi_i]$ is the number of other hooks which intersect the vertical (or horizontal) part of any hook $i \rightarrow \pi_i$, and are represented with green dots.

The dimension of the stock market geometric object is simply:

(II.2)

$$dim(M) = \sum_{i=1}^{n} r[i, \pi_i] - k^2$$

where k is the number of cords in the decorated permutation pointing to the left.

To exemplify, for the permutation $\{2, 4, 5, 7\}$ the number of hook intersections are computed in the figure II.5. Noticing that $k = 2$ this configuration of stock market has the dimension:

(II.3)

$$dim(M) = 7 - 4 = 3$$

In other words it is expected that the geometrical object characterizing the stock market configuration defined by the permutation $\{2, 4, 5, 7\}$ to have 3 dimensions and could be visualize intuitively. It should be notice here that this is a particular case; generally the geometrical object of the stock market lives in higher dimensions and is impossible to be drawn directly.

It was stated in the latter chapters that hooks diagram is an alternative way of finding the Grassmann necklace for a certain stock market configuration. In terms of hooks every term I_i in the Grassmann necklace I is simply a list of k horizontal lines which pass above the ith column. In the example of the permutation $\{2, 4, 5, 7\}$ the Grassmann necklace terms are shown in the figure II.6 below:

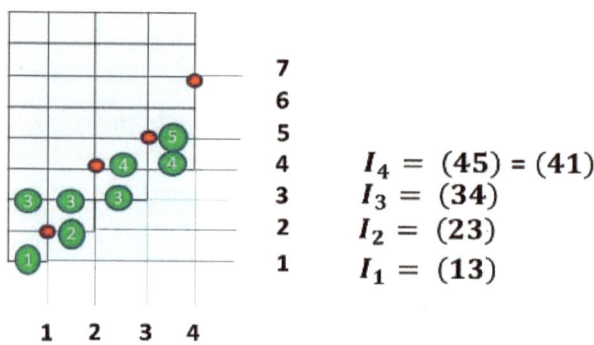

$$I_4 = (45) = (41)$$
$$I_3 = (34)$$
$$I_2 = (23)$$
$$I_1 = (13)$$

Figure II.6. Grassmann necklace resulting from hooks diagram

Notice that that the same Grassmann necklace terms from the last chapter are recovered.

Grassmann necklace is important in the present exposition mainly because it effectively makes the connection to the beautiful

32

structures in combinatorial geometry, such as *matroid, positroid* and *polytopes*, which will be explored in the next chapter.

Chapter 4. Combinatorial geometry and stock market polytope

The following chapter represents mainly a brief introduction to usual combinatorial and algebraic geometry concepts that will be further adapted to stock market realities. The interested reader is referred for extensive exploration of these beautiful branches of mathematics to [3], [4], [14], [15], and [16].

The remarkable combinatorial objects that will be discussed further are the last bricks in the construction of the bridge relating stock market to geometry. The central figures in this section will be matroids , positroid and eventually, polytopes; their relation with the Grassmann necklace will be also explored.

A **matroid** of rank k on the set $\{1, \ldots, n\}$ is a nonempty collection $M \subseteq \binom{n}{k}$ of k elements subsets, called *bases* of M, that satisfies the exchange axiom; for any $I, J \in M$ and $i \in I$, there exist $j \in I$ such that $I \setminus \{i\} \cup \{j\} \in M$.

There is a reference here to the cyclically shifted order that take the mind to Grassmann necklace and a relation that could exist between these combinatorial objects. In fact there is a bijection between Grassmann necklace and a special type of matroid, namely the *positroid*.

Positroid, a relatively new combinatorial concept is mainly the fruit of A. Postnikov work and became "famous "due to its beautiful applications in Quantum Field Theory. The quintessence of Postnikov findings is that each decorated permutation and hence a Grassmann necklace corresponds to a positroid. So to speak every stock market representation, decorated with a permutation of stocks corresponds to a positroid.

A matroid $M \subseteq \binom{n}{k}$ is a **positroid** if and only if it can be written as $SM_{I_1}^1 \cap \ldots \cap SM_{I_n}^n$ for a Grassmann necklace $I = (I_1, \ldots, I_n)$; here $SM_{I_n}^n$ are the cyclically shifted Schubert cells.

Although this is the consecrated definition of positroid in the following, a derivation of it will be used, not only for simplicity, but also for the direct connection with Grassmann necklace.

M is a positroid if and only if the following holds : $H \in M$ if and only if $H \geq_i I_i$ for all $i \in \{1, \ldots, n\}$. In other words the positroid can be written directly from Grassmann necklace by taking:

$$M = \{H | H \geq_1 I_1, \ldots, H \geq_n I_n\}$$

It is clear now why Grassmann necklace is so important.

Turning to the stock market example it will be easy to find the *stock market positroid* out of the decorated permutation associated with the crossing of stocks. Recall that the decorated permutation $\{2, 4, 1, 3\}$ related to state of the stock market has the Grassmann necklace:

(II.4)

$$I_1 = (13), I_2 = (14), I_3 = (23), I_4 = (34)$$

According to the definition, the stock market positroid will have the bases:

(II.5)

$$M = \{(13), (14), (23), (24), (34)\}$$

These abstract "coordinates" reflects the situation of stock market between 5/15/2013 and 6/3/2013, and the stock market positroid encode the crossings of stocks up until this date.

One more step remains until the crossing of stock prices can be reflected in a geometrical object, which as will become immediately clear is a *positroid polytope* or the **stock market polytope**.

The positroid polytope is nothing else than the geometric representation of the positroid having the vertices defined by the positroid bases. In the case of the stock market positroid that has the bases found in (II.5) , the associated *stock market polytope* is graphically depicted in figure II.7:

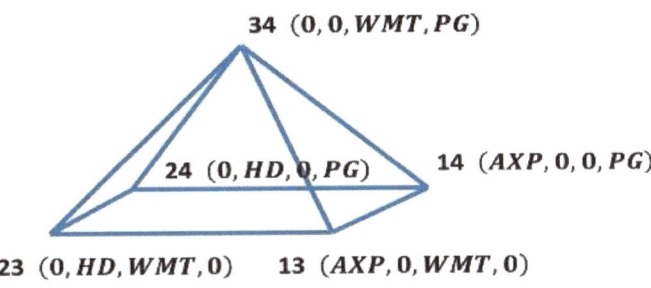

$34 \ (0, 0, WMT, PG)$

$24 \ (0, HD, 0, PG)$ $14 \ (AXP, 0, 0, PG)$

$23 \ (0, HD, WMT, 0)$ $13 \ (AXP, 0, WMT, 0)$

Figure II.7. *The stock market polytope* (stockMarkethedron)

Notice that the polytope vertices are labeled either with numbers and the stocks that are port in the initial stock market permutation.

It is hard to believe that the geometrical construction above could represent the state of the stock market at some specific date, but still the **stock market polytope** encodes all the relevant information related to stocks price moving.

Chapter 5. Taking the Stock Market to higher dimensions

The market configuration of the 4 DJIA components chosen in the example is not the top cell of the Grassmannian G[2,4], as it will become clear in the next chapter. In other words is not the most complex structure the chosen 4 *Dow Jones Industrial Average* can form by crossing their prices paths.

For the sake of completeness, and exemplify for a higher-dimensional stock market polytope, a supplementary crossing of stocks is considered. The added crossing is between stocks 2(HD) and 4(PG), if taken into account the time period between 5/15/2013 and 6/7/2013, such that the resulting stocks crossings diagram can be depicted as:

Figure II.8. The crossing of stocks diagram for the most complex configuration 4 stocks can form

The procedure of finding stock market polytope for any configuration of stock market must be now familiar such that will only be briefly exemplified in what follows.

The permutation associated to the diagram in figure II.8 is:

$$\{3, 4, 1, 2\},$$

and the permutation diagram could be constructed as:

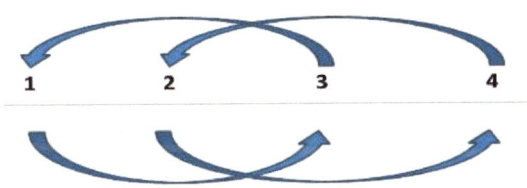

Figure II.9. Permutation diagram for the stock market configuration in figure II.8

The permutation diagram is included here only for didactical reasons, to see how it differ from the diagram associated to the 4 DJIA components having 3 crossings, that was explained in chapter 2. For defining the Grassmann necklace of the stock market configuration related to permutation$\{3, 4, 1, 2\}$, and the dimensions stock market polytope will have, the hook diagrams are directly employed.

To conclude on the dimensions of the polytope connected to Dow Jones Industrial Average configuration from 5/15/2013 to 6/7/2013, the hooks in figure II.10 is drown.

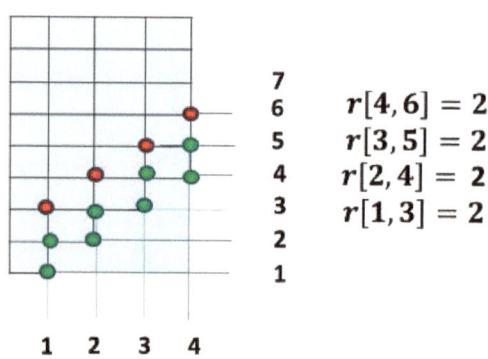

$r[4,6] = 2$
$r[3,5] = 2$
$r[2,4] = 2$
$r[1,3] = 2$

Figure II.10. Hooks diagram associated to permutation $\{3, 4, 1, 2\}$

A quick computation of dimensions the stock market polytope will have leads to:

$$dim(M) = 8 - 4 = 4$$

This time the **stockmarkethedron** has higher dimensions (respectively 4) and its shape cannot be intuitively visualized.

The next step is to find the Grassmann necklace and its associated matroid (positroid) basis. The hook diagram in figure II.11 will help realize this objective.

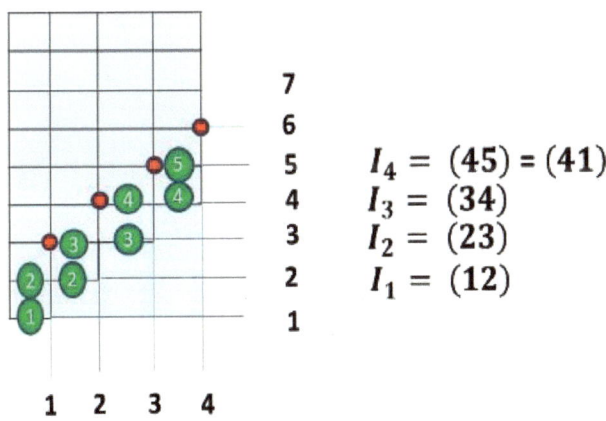

$$I_4 = (45) = (41)$$
$$I_3 = (34)$$
$$I_2 = (23)$$
$$I_1 = (12)$$

Figure II.11. Grassmann necklace resulting from hooks diagram related to permutation $\{3, 4, 1, 2\}$

As it can be easily seen the Grassmann necklace is:

(II.6)

$$I_1 = (12), I_2 = (14), I_3 = (23), I_4 = (34)$$

According to the chapter 4 definition, the stock market positroid will have this time the bases:

(II.7)

$$M = \{(12),\ (13),\ (14),\ (23),\ (24),\ (34)\}$$

Although is not possible to visualize the polytope having the above basis as vertices, figure II.12 sketched a geometrical shape in 4-dimensions that could (really?) give a hint of the complexity of the multidimensional geometry.

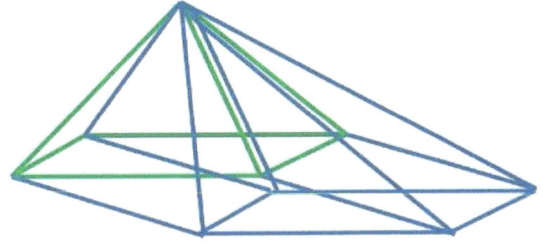

Figure II.12. A sketch of a 4-dimensional geometrical shape

I will emphasize here that the figure above is NOT the stockmarkethedron connected to positroid having the basis deduced in (II.7). The solely porpoise of the geometrical construction in figure II.12 is to help the reader visualizing 4-dimensional shapes.

It should be noted here that the last chapter stock market polytope, the one associated to 3 crossings of the 4 DJIA stock components will be **a face** of the higher dimensional polytope resulted for accounting of the 4 crossings. One of the faces of the polytope in figure II.12 is drown in green to be easily distinguished.

The **added crossing of stocks changes the shape of the stock market polytope**. The shape of the stock market polytope is changing every time a price quotation added a new crossing of stocks.

Chapter 6. The Positive Grassmannian and the decomposition of the stock market polytope

An alternative and more fruitful way to construct the stock market polytopes is to consider the matrix approach of Grassmannians variety. The present paper intentionally avoids this approach due to its complexity and more deeply implications that will constitute the subject of a future exposition. The richness of the matrix in approaching the stock market will reside eventually in expressing the probability of finding the market in a certain state only by computing the volume of the stock market polytope.

Here the discussion is limited to showing the relation between the stock market polytope and the remarkable structures of ***positive Grassmanians*** that leads to stratification and decomposition into positroid cells. Simply speaking the decomposition of stock market polytope is nothing else but the elimination of stock crossings one of a time. The inverse operation, namely amalgamation is related to construction of the stock market polytope by adding crossing of stocks to the initial configuration of the market.

The Grassmannian $G(k, n)$ is the space of k-dimensional planes in n-dimensional space. The positive Grassmannian $G^+(k, n)$ is a subset of the real Grassmannian $G(k, n)$. In [32], [33] and [34] Postnikov shows that a positroid is a point in the ***positive Grassmannian***, indexed by the Grassmann necklace. Also a positroid is associated to a *positive Grassmann cell* such as the $G^+(k, n)$ is the disjoint union of its cells.

It looks complex but as it will be seen immediately for the stock market example decorated by the permutation $\{2, 4, 1, 3\}$ that stock market polytope is simply the union of all the stock crossings. Removing the crossings one of a time reside in decomposition of the positive Grassmannian.

In order to better visualize the details of the decomposition, the permutation diagram of the stocks is used once again. This time the permutation diagram is associated to another combinatorial object, the **non-crossing partition** in the way shown in figure II.13.

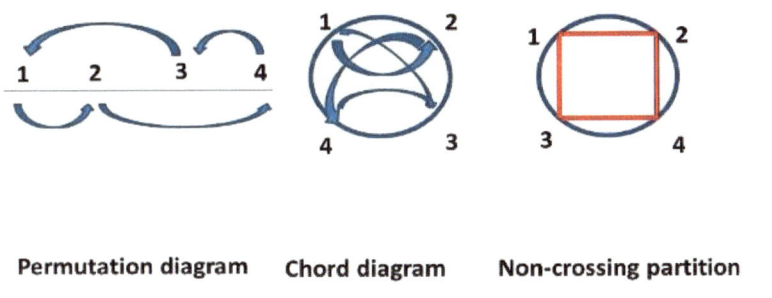

Permutation diagram   **Chord diagram**   **Non-crossing partition**

Figure II.13. Non-crossing partition of the stocks decorated permutation

The particularity of non-crossing partition is that its edges do not intersect. This is exactly what happens if a cross of stocks is removed from the initial permutation.

In the non-crossing partition in figure II.13 removing the edge (34) it follows that the edge (24) vanish too, otherwise the non-crossing partition no longer exist. The situation is explicit in figure II.14.

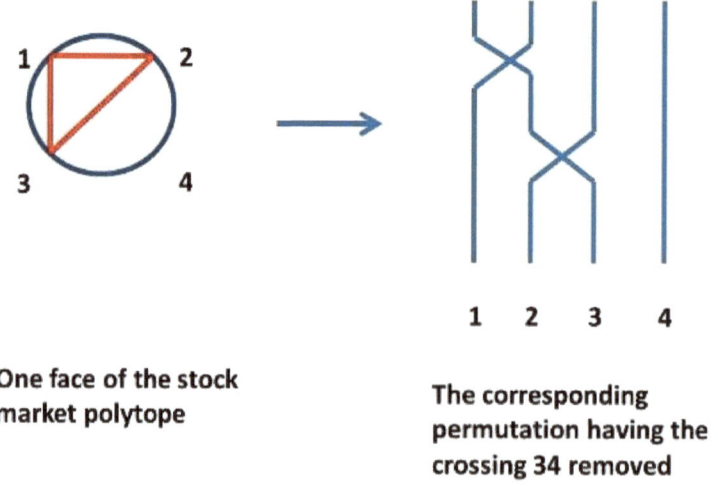

One face of the stock market polytope

The corresponding permutation having the crossing 34 removed

Figure II.14. Removal of stock crossing associated with a face of stock market polytope

It is clear now from the wiring diagram that removing (34) cancels the (24) in the same time. This cancelation in pairs preserves the positivity of the Grassmannian $G^+(2, 4)$ after each crossing removal. Each face of a positroid is a positroid.

Removals of crossings in the permutation associated to stock market reside in the decomposition of the positive Grassmannian $G^+(2, 4)$ labeled by the stock market polytope in positroid cells as is depicted in the figure II.15.

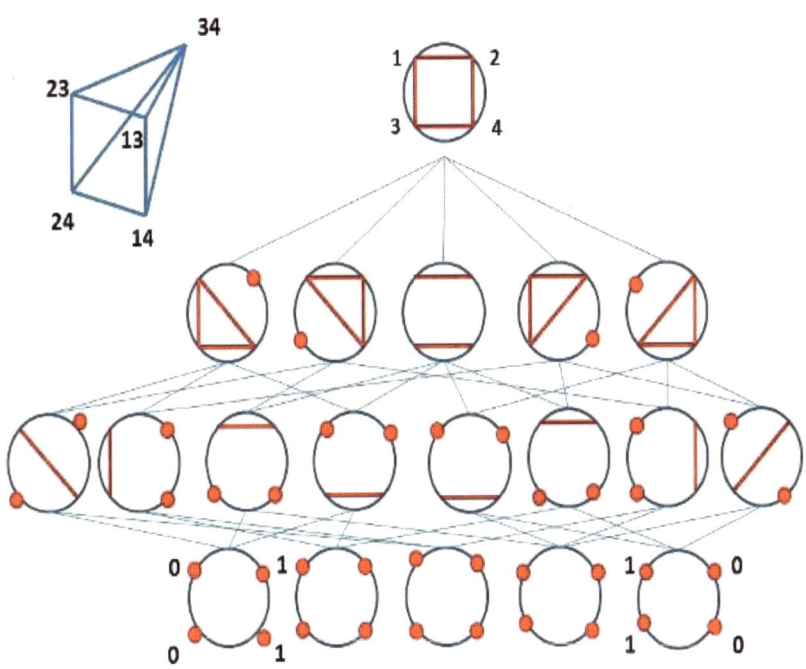

Figure II.15. Decomposition of the stock market polytope

Notice that every crossing corresponds to a face of the polytope. The stock market polytope can be constructed step-by-step from

47

the initial state of the market by adding edges and faces every time a market quotation induces a stock crossing. Adding positroid cells is called the gluing procedure.

Decomposition and the gluing procedure are more complex when all the *Dow Jones Industrial Average* stock components are considered. An increasing number of stock crossings have to be taken into account, a procedure that leads to a space having higher dimensions.

REMARKS: Prof. Allen Knutson has remarked that the decomposition procedure of the stock market polytope can lead to cells being in different Grassmanians. It seems likely that the present scheme might be only defined for permutation. The issue remains open for now.

Conclusions of the section

The present section constitutes a map connecting the price evolution of stocks in the market with complex and beautiful geometrical shapes. The "roads" of this map are combinatorial concepts as permutations and hooks, positroid and polytopes or the remarkable geometric structures of positive Grassmannians.

The key point in constructing the above mentioned map is to associate changings in stocks prices with combinatorial objects such as permutations. This association is easy to make if the stocks that compose an index are arranged in a particular way (explained in section I) and take into account the crossing of stocks. Under some conventions the diagram that results by winding the prices of stocks is exactly a wiring diagram associated to permutations.

To exemplify with real stock quotations, *Dow Jones Industrial Average* (DJIA) index is analyzed and prices of four of its stock components (APX, HD, WMT and PG) for a short period of time are integrated in a wiring diagram that is proved to be a remarkable permutation diagram. In other words it can be said about the stocks in discussion that AXP permutes with HD, or that WMT permutes with PG.

Pushing further the combinatorial approach of the stock market, the resulting wiring diagram is transformed according to some simple rules in a permutation diagram, with the help of decorated permutation. Some combinatorial objects related to *decorated permutation* are briefly explained, along with their connections with stock market, without diving too deep into the associated mathematics.

Combinatorial notions, like Grassmann necklace, matroid or positroid could look exotic in the financial frame, but as it was seen in the present exposition they fit well in the picture of stock market.

 Finally the map connecting the price quotations of stocks in the market with beautiful geometrical shapes is completed by adding the concept of polytopes. This way the geometrical object that encodes all the relevant information of stock market can be named as **stock market polytope**.

To better understand the polytopes, the *positive Grassmannians* are added to complete the picture in which the performances of stock market can be viewed by geometrical shapes.

 Since typically polytopes live in higher dimensions their shapes cannot have a direct image. The positive Grassmannian cells help in the construction of multidimensional stock market polytopes by simply gluing together all the stock crossings between components of the market index.

Certainly is not easy to imagine the stock market as a geometric shape, such as a pyramid to name a trivial example. Still the stock market polytope fully encodes all the relevant information about the current (or future as a future paper will assess) state of the stock market.

Deeper insights into the properties of stock market polytopes such as the relation between its volume and the probability to a certain state of stock market to occur are exploring in future research papers.

Bibliography

[1] Amariti A. and Forcella D., Scattering Amplitudes and Toric Geometry, 2013,
arXiv:1305.5252.

[2] Ardila F., Rincon F., Williams L., Positroids and non-crossing partitions, 2013, arXiv:1308.2698.

[3] Arkani-Hamed N., Bourjaily J. L., Cachazo F., Goncharov A. B., Postnikov A., and Trnka J., Scattering Amplitudes and the Positive Grassmannian, 2012, arXiv:1212.5605.

[4] Arkani-Hamed N., Bourjaily J. L., Cachazo F., Hodges A., and Trnka J., A Note on Polytopes for Scattering Amplitudes," JHEP 1204 (2012) 081, arXiv:1012.6030 [hep-th]. 4, 6, 11, 56, 57, 69, 248.

[5] Black F., Jensen M.C., Scholes M., Studies in the Theory of Capital Markets, edited by M. Jensen New York: Praeger Publishers, 1972.

[6] Campbell J., Lo A., MacKinley A., The Econometrics of Financial Markets, Princeton University Press, 1997.

[7] Gelfand I. M., Goresky R. M., MacPherson R. D., and Serganova V. V., Combinatorial Geometries, Convex Polyhedra, and Schubert Cells," Adv. In Math. 63 (1987) no. 3, 301{316. 151}.

[8] Gottman, J. M., Time Series Analysis – A Comprehensive Introduction for Social Scientists, Cambridge University Press, 1981.

[9] Fama E., The Behavior of Stock Market Prices, J. Business 38, pp. 34-105, 1965.

[10] Franco S., Galloni D., and Mariotti A., The Geometry of On-Shell Diagrams, 2013,
arXiv:1310.3820.

[11] Kenett D.Y., Preis T., Gur-Gerschgoren G., Ben-Jacob E., Quantifying meta-correlations in financial markets, 2012, EPL 99, 38001.

[12] Kodama Y. and Williams L., KP Solitons and Total Positivity for the Grassmannian, arXiv:1106.0023 [math.CO]. 245

[13] Postnikov A., Total positivity, Grassmannians, and networks, arXiv Mathematics e-prints (2006) [math/0609764].

[14] Postnikov A., Speyer D., and Williams L., Matching polytopes, toric geometry, and the non-negative part of the Grassmannian, Journal of Algebraic Combinatorics 30 (2009)173-191, [arXiv:0706.2501].

[15] Postnikov A., Positive Grassmannian,
http://www-math.mit.edu/~ahmorales/18.318lecs/lectures.pdf, 2013.

[16] Racorean O., Crossing of Stocks and the Positive Grassmannian I : The Geometry behind Stock Market,
http://arxiv.org/abs/1402.1281, 2014.

[17] Suho Oh, Positroids and Schubert matroids, Journal of Combinatorial Theory, Series A 118 (2011), no. 8, 2426-2435, arXiv:0803.1018 [math.CO].

[18] Weigend A.S., Gershenfeld N.A. , Time series Prediction: Forecasting the Future and Understanding the Past, Reading, MA: Addison Wesley, 1994.

www.ingramcontent.com/pod-product-compliance
Lightning Source LLC
Chambersburg PA
CBHW040850180526
45159CB00001B/380